American Institute of Chemical Engineers

Environmental Division Conference

Presentations at the
2008 AIChE Spring National Meeting

April 6-10, 2008
New Orleans, Louisiana, USA

Printed from e-media with permission by:

Curran Associates, Inc.
57 Morehouse Lane
Red Hook, NY 12571
www.proceedings.com

ISBN: 978-1-60560-219-6

Some format issues inherent in the e-media version may also appear in this print version.

ISBN: 978-1-60560-219-6

American Institute of Chemical Engineers
Proceedings
Three Park Avenue
New York, NY 10016-5991
Phone: 212-591-8100

www.aiche.org

American Institute of Chemical Engineers

Environmental Division Conference
2008

TABLE OF CONTENTS

TREATMENT AND COLOR REMOVAL OF WASTEWATERS USING WET AIR OXIDATION

Bryan Kumfer and Tim Schleusner
Siemens Water Technologies
301 W. Military Rd
Rothschild, WI 54474

ABSTRACT

Wet air oxidation (WAO) is a technology used to treat high strength wastewaters that typically have a chemical oxygen demand of 10,000 mg/L to 500,000 mg/L. These wastewaters typically contain compounds that produce a highly colored wastewater. Often, this color cannot be removed by conventional biological processes. Bench scale testing has shown that WAO is effective for the color removal from dye wastewaters. Greater than 98% of the color is removed while reducing the COD by more than 60% while producing a readily biodegradable effluent. WAO is also effective in the color removal from molasses type wastewaters. Color removal of greater than 99% was achieved with molasses in deionized water. In addition, a molasses based distillery wastewater and an anaerobically treated distillery wastewater showed a significant color reduction after WAO treatment while achieving COD reductions of 86% and 80%, respectively and producing a readily biodegradable WAO effluent. Greater than 97% color removal was achieved by WAO of a wastewater from acrylonitrile production. This allowed for white ammonium sulfate crystals with a low TOC concentration to be recovered during crystallization of the WAO effluent.

INTRODUCTION

Industrial processes can produce highly colorized wastewaters that contain a high chemical oxygen demand (COD) concentration. These wastewaters often contain compounds that are bio refractory or even toxic to biological treatment systems. Wastewaters from distilleries, wastewaters containing dyes, and wastewaters from acrylonitrile production are three examples of wastewaters that contain refractory compounds that are highly colorized.

Dyes are common in wastewaters from textile industries. Conventional biological treatment does not effectively remove the color from these wastewaters. Ion exchange and sorbents, such as zeolites and clays, have been shown to remove the color compounds from the wastewater.[1] However, these types of treatments still require additional treatment or disposal of the sorbents containing the dye compounds. Oxidation technologies, such as Fenton's reagent, have been shown to be effective at removing color, but these technologies are typically most effective with low COD wastewaters.[2]

Molasses based wastewaters, such as from distilleries, typically have a very high COD (50,000 – 200,000 mg/L) and are biodegradable. Even though the wastewaters are biodegradable, there are colored compounds that are refractory. These wastewaters still have a dark brown color after aerobic or anaerobic treatment.[3] These compounds are not able to be removed with simple absorption with activated carbon. Color removal has been accomplished by using membrane technologies, such as RO, but additional treatment or disposal of the concentrated wastewater is still required. Ozone has been shown to be effective for removing color from distillery wastewaters, but due to the very high COD concentration, this process is not cost effective.[4]

During the production of acrylonitrile, a waste stream consisting of a high concentration of ammonium sulfate with additional organics is produced. Often, the ammonium sulfate is recovered by a crystallization process. However, the blow down stream from the crystallization process contains a high COD concentration and requires further treatment. The wastewater often contains inhibitory chemicals that prohibit treatment by biological processes[5]. If the organics in this blow down stream are removed or destroyed, the ammonium sulfate in the blow down stream can be recovered as a usable product.

The wet air oxidation (WAO) technology is a process that is used to oxidze color compounds, reduce chemical oxygen demand and increase biotreatability of wastewaters. WAO is often used to treat high COD (10,000 – 500,000 mg/L) wastewaters that often have limited biodegradability. WAO uses temperature, pressure, and an oxygen source, usually air, to treat the wastewaters. WAO is typically operated in the temperature range of 120 °C to 320 °C and pressures of 100 psi to 4500 psi. During the WAO process, the complex color causing organic compounds are converted to carbon dioxide, water, and biodegradable short chain organic acids, such as acetic acid and oxalic acid.[6] Inorganic compounds, such as sulfides, are converted to non-reactive inorganic compounds. Following WAO treatment, the effluent is typically pale yellow to near colorless and any remaining COD can be treated by conventional biological treatment.

EXPIREMENTAL PROCEDURE

The WAO tests were performed in laboratory autoclaves. An illustration of the bench scale shaking autoclaves used in the tests is shown in Figure 1. The autoclaves were charged with the test solution, catalyst or oxidation enhancer if needed, and compressed air. The charged autoclaves were placed in a heater/shaker unit and heated to the desired temperature. The autoclaves were held at temperature for the reaction time, removed from the heater, and cooled to room temperature using tap water.

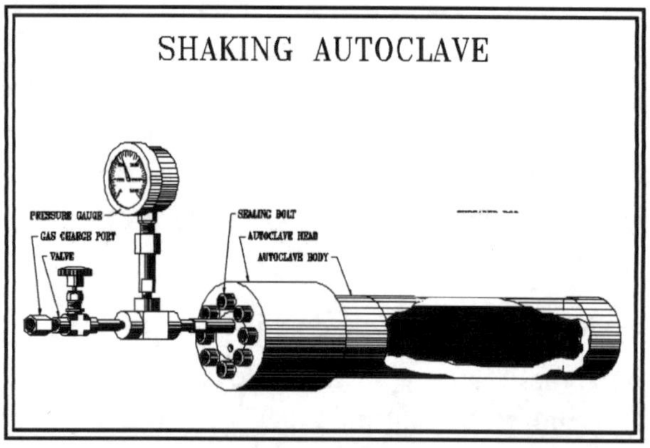

Figure 1. Laboratory shaking autoclave

DISCUSSION AND RESULTS

AZO Dye

The test solution consisted of a 5 g/L indigo dye in deionized water. The solution was oxidized at 200, 240 and 280 °C with a 60 minute time at temperature. The indigo dye solution and WAO effluents are presented in Figure 2. The results from WAO testing are presented in Figures 3 and 4.

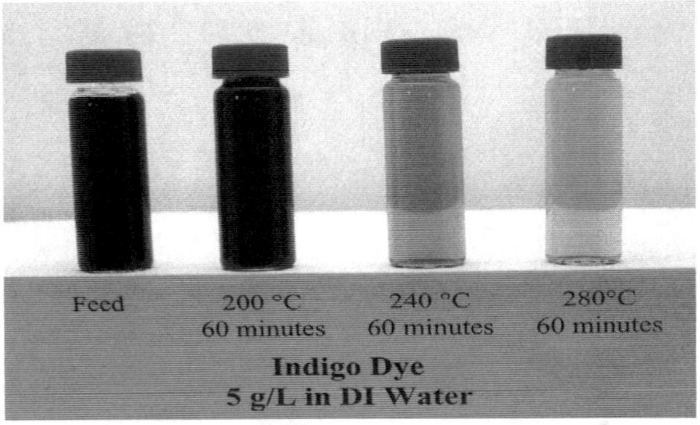

Figure 2. Indigo feed and WAO effluents.

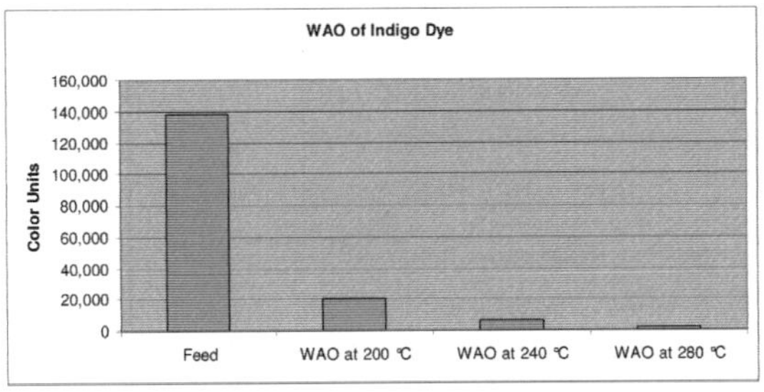

Figure 3. Color results for indigo feed and WAO effluents

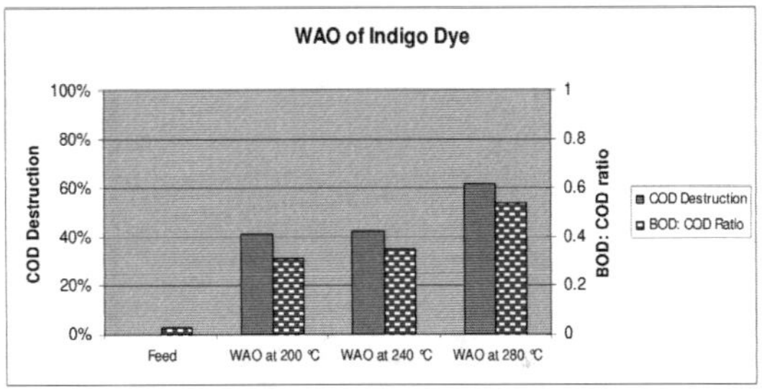

Figure 4. COD destruction and BOD: COD ratios for indigo feed and WAO effluents.

The color of the solution decreased greater than 85% when oxidized at 200 °C with a 60 minute time at temperature. As the oxidation temperature increased to 280 °C, the color removal achieved was greater than 98%. In addition, as the temperature increased, the percent COD destruction increased and the biodegradability of the WAO effluent increased. The feed solution had a BOD: COD ratio of 0.03 which indicates that it is not biodegradable. At 200 °C, the BOD: COD ratio increased to 0.31 which indicates that the effluent is more biodegradable. At 280 °C, the BOD: COD ratio of the effluent was greater than 0.5. A BOD: COD ratio of 0.5 or greater typically indicates a readily biodegradable effluent.

Molasses Solution

The test solution consisted of 30 g/L molasses in deionized water. The solution was oxidized at 200, 240 and 280 °C with a 60 minute time at temperature. The molasses feed and WAO effluents are presented in Figure 5. The results from WAO testing are presented in Figures 6 and 7.

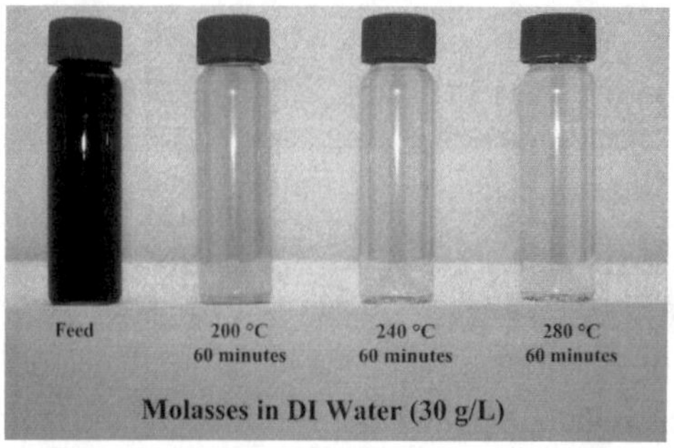

Figure 5. Molasses feed and WAO effluents.

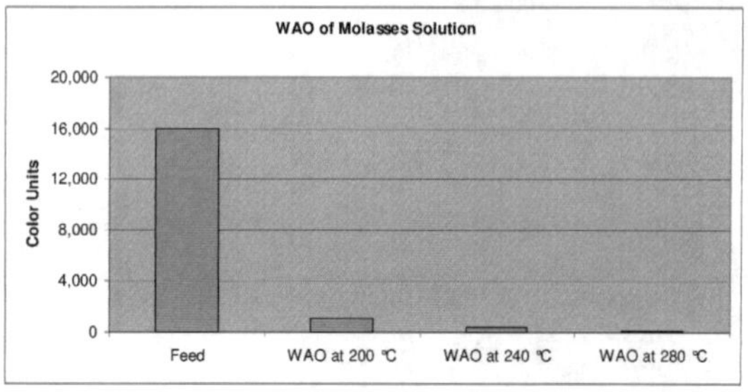

Figure 6. Color results for molasses feed and WAO effluents

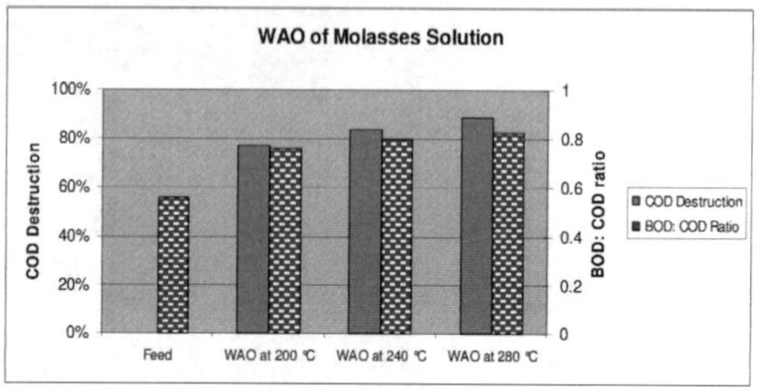

Figure 7. COD destruction and BOD: COD ratios for molasses feed and WAO effluents.

The color of the solution decreased by more than 93% when oxidized at 200 °C with a 60 minute time at temperature. As the oxidation temperature increased to 280 °C, the color removal achieved was greater than 99.5%. Again, as the oxidation temperature increased, the percent COD destruction increased. The feed solution had a BOD: COD ratio of greater than 0.5, which indicates that it is readily biodegradable. However, when the solution is WAO, the BOD: COD ratio increases to greater than 0.75, indicating that the biodegradability of the solution increased.

Distillery Waste Water

The distillery wastewater that was tested had a feed COD concentration of 92 g/L and a BOD: COD ratio of 0.38. Due to the concentrated COD demand of the feed, a 1:2 (1 part feed: 1 part DI water) dilution was performed for bench scale testing. The diluted feed was oxidized at 240, 260, and 280 °C with a 60 minute time at temperature. Two additional oxidations were performed at 280 °C with a 60 minute time at temperature using nitric acid as an oxidation enhancer. The distillery wastewater and WAO effluents are presented in Figure 8. The destruction of COD and the resulting BOD: COD ratio for the oxidations is presented in Figure 9.

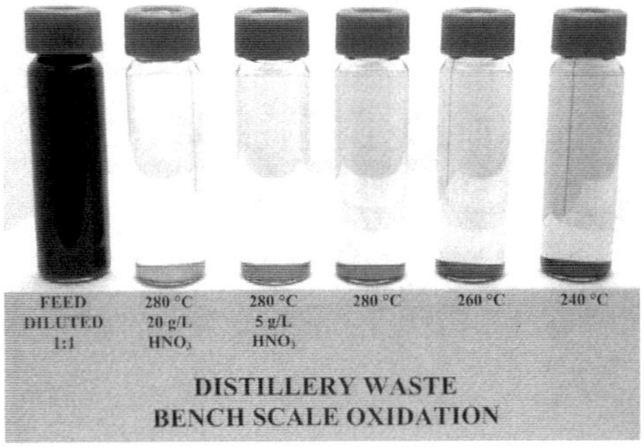

Figure 8. Distillery wastewater and WAO effluents.

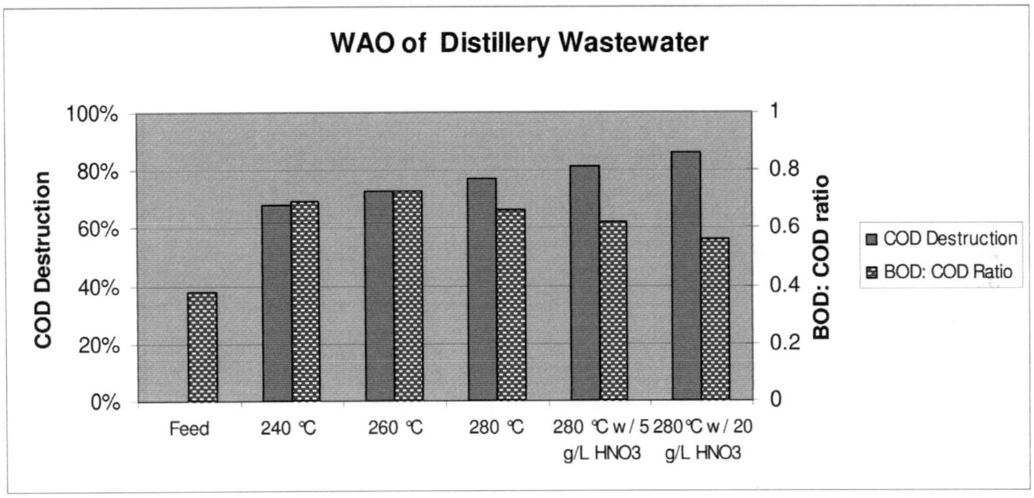

Figure 9. COD destruction and BOD: COD ratios for distillery wastewater and WAO effluents.

There was no color analysis performed on the WAO effluents. However, a visual observation shows that the color of the oxidized effluents ranged from a clear yellow color to colorless which indicates a high removal of color. In general, as the temperature increased and the concentration of nitric acid increased, the residual color decreased. The effluent from the 240 °C oxidation had high color removal, but the 280 °C with 20 g/L HNO_3 effluent was nearly colorless. All WAO effluents had a BOD: COD ratio greater than 0.5, which indicates that the effluent is readily biodegradable. The effect of WAO on COD destruction follows the same trend as color removal. As the temperature and concentration of nitric acid increased, the COD destruction achieved increased. The COD destruction ranged from 68% at 240 °C without nitric acid added to 86% at 280 °C with 20 g/L HNO_3 added.

Digested Distillery Waste

An anaerobically digested distillery waste was tested for color removal. The wastewater had a COD concentration of 48 g/L and a BOD: COD ratio of less than 0.2. A series of oxidations were performed on the feed solution at temperatures ranging from 220 °C to 280 °C with a time at temperature between 30 and 90 minutes. The anaerobically digested distillery wastewater and WAO effluents are presented in Figure 10. The destruction of COD and the resulting BOD: COD ratio for the oxidations is presented in Figure 11.

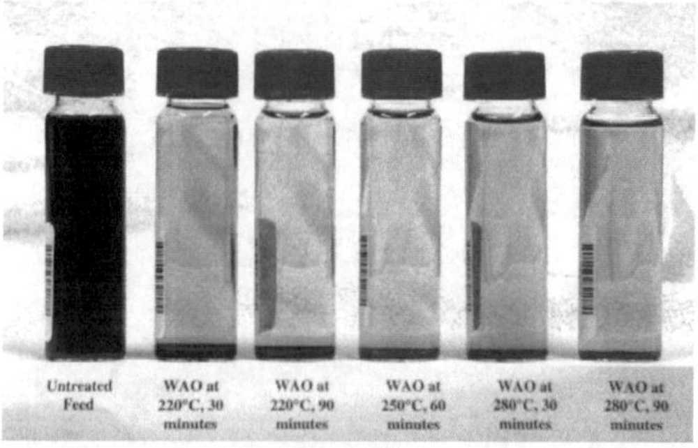

Figure 10. Anaerobically digested distillery wastewater and WAO effluents.

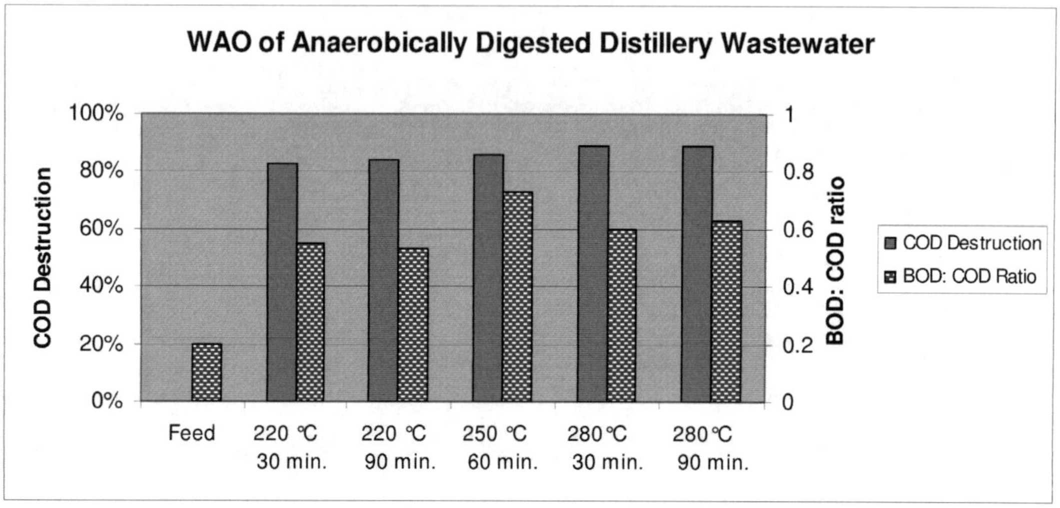

Figure 11. COD destruction and BOD: COD ratios for anaerobically digested distillery wastewater and WAO effluents.

There was no color analysis performed on the WAO effluents. However, a visual observation shows that all conditions had a clear yellow color after WAO which indicates a high removal of color. All WAO effluents had a BOD: COD ratio of 0.5 or higher, which indicates a readily biodegradable effluent. Since the feed had a BOD: COD ratio of less than 0.2, this is a significant improvement in biodegradability. The COD destruction for all conditions was greater than 80%. Again, as the temperature and time at temperature was increased, the COD destruction achieved increased.

Acrylonitrile Wastewater

An ammonium sulfate solution from the production of acrylonitrile was tested for color removal. The feed solution had a COD of 204,000 mg/L. Due to the high COD, the wastewater was diluted 1:2 (1 part wastewater: 1 part deionized water) for bench scale testing. For this testing, the WAO effluent was to be crystallized to recover the ammonia sulfate so the biodegradability of the wastewater was not obtained. The acrylonitrile wastewater and WAO effluents are presented in Figure 12. The color removal and the destruction of COD for the oxidations are presented in Figure 13.

Figure 12. Acrylonitrile wastewater and WAO effluents.

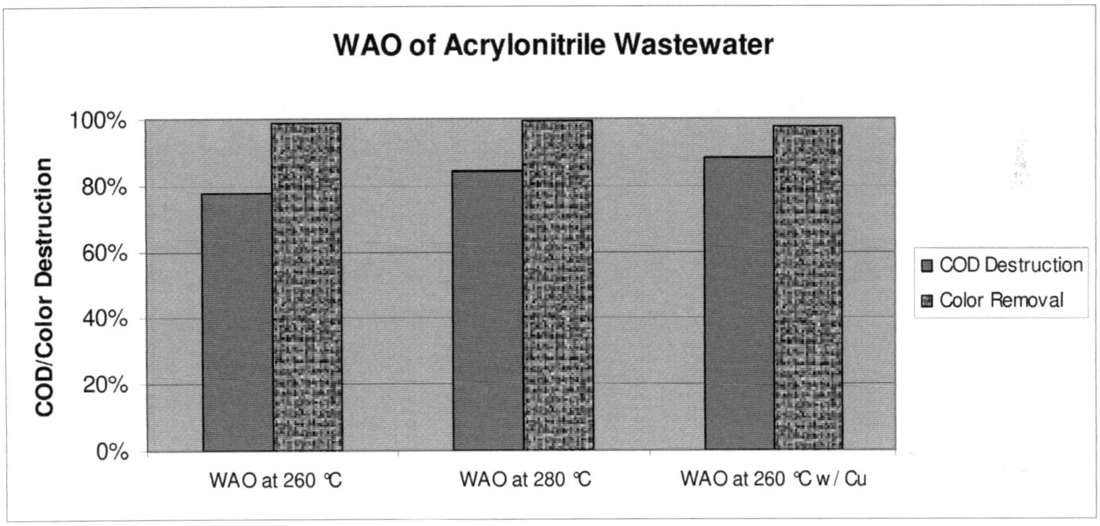

Figure 13. COD destruction and color removal for the WAO effluents.

The color of the solution decreased greater than 98% when oxidized at 260 °C with a 60 minute time at temperature. As the oxidation temperature increased to 280 °C, the color removal achieved was greater than 99.5%. When copper was added as a catalyst, the color removal decreased slightly to 97.7%, but this is most likely due to the soluble copper in solution. Again, as the temperature increased, the percent COD destruction increased as well. With the use of the catalyst, the COD destruction was even greater. The COD destruction ranged from 77% without the use of catalyst to 89% with the use of catalyst. A bench scale crystallization test was performed with the effluent from the WAO at 260 °C and the use of copper catalyst. The ammonium sulfate crystals formed only contained 1060 mg/kg TOC. As shown in Figure 14, the recovered ammonium sulfate crystals are white in color.

Figure 14. Recovered ammonium sulfate crystals from the WAO effluent.

CONCLUSIONS

WAO is effective for the color removal from dye wastewaters. Greater than 98% of the color is removed while reducing the COD by more than 60%. In addition, the WAO effluent shows a significant improvement in biodegradability. WAO is also effective in the color removal from molasses type wastewaters. Color removal of greater than 99% was achieved with molasses in deionized water. In addition, a molasses based distillery wastewater and an anaerobically treated distillery wastewater showed a significant color reduction after WAO treatment. A COD destruction of greater than 80% was achieved for both types of wastewaters while producing a readily biodegradable WAO effluent. Greater than 97% color removal was achieved by the WAO of a wastewater from acrylonitrile production. The crystallization of the WAO effluent produced white crystals with a low TOC concentration.

REFERENCES

1. Maheria, K.; Chudasama, U. Sorptive Removal of Dyes Using Titanium Phosphate. *Ind. Eng. Chem. Res.* **2007**, *46*, 6852-6857.
2. Herney-Ramirez, J.; lampinen, M.; Vicente, M.; Costa, C.; Madeira, L. Experimental Design to Optimize the Oxidation of Orange II Dye Solution Using a Clay-based Fenton-like Catalyst. *Ind. Eng. Chem. Res.* **2008**, *47*, 284-294.
3. Satyawali, Y.; Balakrishnan, M. Wastewater treatment in molasses-based alcohol distilleries for COD and color removal: A review. *J. Environ. Manag.* **2008**, *86*, 481-497.
4. Sreethawong, T.; Chavadej, S.; Color Removal of distillery wastewater by ozonation in the absence and presence of immobilized iron oxide catalyst. *J. Hazard. Mater.* **2008**, doi:10.1016/j.jhazmat.2007.11.091
5. Kumar,A.; Prasad, B.; Mishra, I.; Optimization of process parameters for acrylonitirile removal by a low-cost adsorbent using Box-Behnken design. *J. Hazard. Mater.* **2008,** *150*, 174-182
6. Maugans, C.; Ellis, C. "Wet Air Oxidation: A review of Commercial Sub-Critical Hydrothermal Treatment", International Conference on Incineration and Thermal Treatment Technologies (IT3), New Orleans, LA, May 13-17, 2002.

The disposal of neutralized chemical warfare materials using a commercial wet oxidation system: A case study

By:
Clay Maugans – Siemens Water Technologies
Bill Adams – U.S. Army Non-Stockpile Chemical Materiel Project
Ed Doyle – U.S. Army Non-Stockpile Chemical Materiel Project
Jeff Bettinger – Shaw Environmental, Inc.
Bill Killilea – Shaw Environmental, Inc.
Kevin Laymon – Shaw Environmental, Inc.

Abstract:
In order to meet the requirements of an International Treaty mandating the destruction of chemical weapons, the US Army in 2005-06 neutralized its stockpile of DF and QL binary chemicals which are used to make GB and VX nerve agents. In order to completely meet Treaty requirements, this neutralized wastewater must be further treated to destroy Schedule 2 chemical breakdown products.

Wet Oxidation was selected to provide this additional treatment. A commercial system was installed in 2006 and operations began in 2007. All of the QL neutralent wastewater destruction was completed in April, 2007. The DF neutralent destruction was completed in November, 2007. This paper describes the wet oxidation system and its performance.

Introduction:
The United States is a signatory on the Chemical Weapons Convention, which requires the disposal of chemical warfare material. The treaty requires that all chemical weapons must be destroyed and that certain residual compounds, regardless of how common or benign such compounds are, must also be destroyed to prove complete destruction. One such class of residual compounds is material containing a carbon-phosphorus (C-P) chemical bond.

The U S Army Non-Stockpile Chemical Materiel Project (NSCMP) is responsible for the disposal of certain chemical warfare materials. The NSCMP contracted with Shaw Environmental to investigate and pursue non-incineration based technologies for this application. From 2001 to 2003, Shaw contracted Siemens Water Technologies (formerly USFilter/Zimpro) in Rothschild, Wisconsin to evaluate the use of Zimpro® wet air oxidation (WAO) to treat these materials (Maugans et al., 2004).

Based on the results of that evaluation, a full-scale WAO system was constructed in Houston, Texas at a commercial hazardous waste Treatment, Storage, and Disposal Facility (TSDF). The treatment facility was commissioned in 2006 and began processing NSCMP materials in 2007.

WAO is a liquid phase reaction in water using dissolved oxygen to oxidize wastewater contaminants. The oxidation reactions occur at moderate temperatures of 150 to 320 °C and at pressures from 10 to 220 bar. The operating pressure maintains water in the liquid phase, which is necessary in order for the oxidation reactions to occur. The pressure also helps regulate the temperature. Energy consumption is minimized by the high pressure operation, since this prevents the energy intensive boiling and vaporization of the water. The process can convert organic contaminants to carbon dioxide, water, and biodegradable short chain organic acids. Inorganic constituents such as sulfides and cyanides are also oxidized.

In the WAO reaction, complex organic molecules are broken into simpler organic compounds, or to a complete mineralized state (CO_2, NH_3, F^-, PO_4^{-3}). In practice, the complex organic feeds are generally not completely mineralized. Simple organic compounds such as low molecular weight carboxylic acids and mineralized reaction products will be present in the process effluents. WAO off-gas has negligible particulates, NO_x, and SO_x but can contain some volatile organic compounds (VOC). To treat the VOC, further off-gas treatment is sometimes required using conventional technologies such as a regenerative thermal oxidizer (RTO).

Feed Description

The neutralents were made by mixing the DF or QL with water. This was done on-location by the Army at Pine Bluff Arsenal, AR. The neutralents were then trucked to the TSDF in Houston and were stored on site in storage tanks. The neutralents require different operating concentrations from each other so are processed separately. The composition of the neutralents is shown in Tables 1 and 2.

Table 1. DF Neutralent Composition

Component	Formula	Name	wt%
MPA	CH3PO(OH)2	Methylphosphonic Acid	20.58%
HF	HF	Hydrogen Fluoride	8.43%
MP	CH3PO(H)(OH)	Methylphosphinic Acid	0.00%
KB	HOCH2CH2N(CHC2H6)2	Diisopropylamino-ethanol	0.00%
ZS	CH3CH2OH	Ethanol	0.00%
NaF	NaF	Sodium Fluoride	0.30%
NaOH	NaOH	Sodium Hydroxide	0.00%
H2O	H2O	Water	70.69%
		Totals	100.00%
		gallons	153,515
		density	1,120 g/L
		fluoride concentration to WAO	2.000 g/L
		dilution req'd (WAO/Raw Feed)	45.59
		diluted COD (th) (WAO feed)	3.370 g/L
		diluted KOH for F, P in feed(th) (WAO feed)	18.60 g/L

Table 2. QL Neutralent Composition

Component	Formula	Name	wt%	
MPA	CH3PO(OH)2	Methylphosphonic Acid	0.00%	
HF	HF	Hydrogen Fluoride	0.00%	
MP	CH3PO(H)(OH)	Methylphosphinic Acid	5.24%	
KB	HOCH2CH2N(CHC2H6)2	Diisopropylamino-ethanol	9.50%	
ZS	CH3CH2OH	Ethanol	3.01%	
NaF	NaF	Sodium Fluoride	0.00%	
NaOH	NaOH	Sodium Hydroxide	0.25%	
H2O	H2O	Water	82.00%	
		Totals	100.00%	
		gallons	74,810	
		density	1,010	g/L
		max COD concentration to WAO	23.200	g/L
		dilution req'd (WAO/Raw Feed)	15.49	
		diluted COD (th) (WAO feed)	23.200	g/L
		diluted KOH for F, P in feed(th) (WAO feed)	41.40	g/L

Description of the WAO System

The purpose of the WAO system is to destroy the carbon-phosphorus chemical bonds (C-P bonds) in the neutralents. A photograph of the system is shown in Figure 1. A more detailed photograph of the oxidation equipment is shown in Figure 2. The process flow diagram is shown in Figure 3.

Figure 1. NSCMP Treatment System

Figure 2. WAO System.

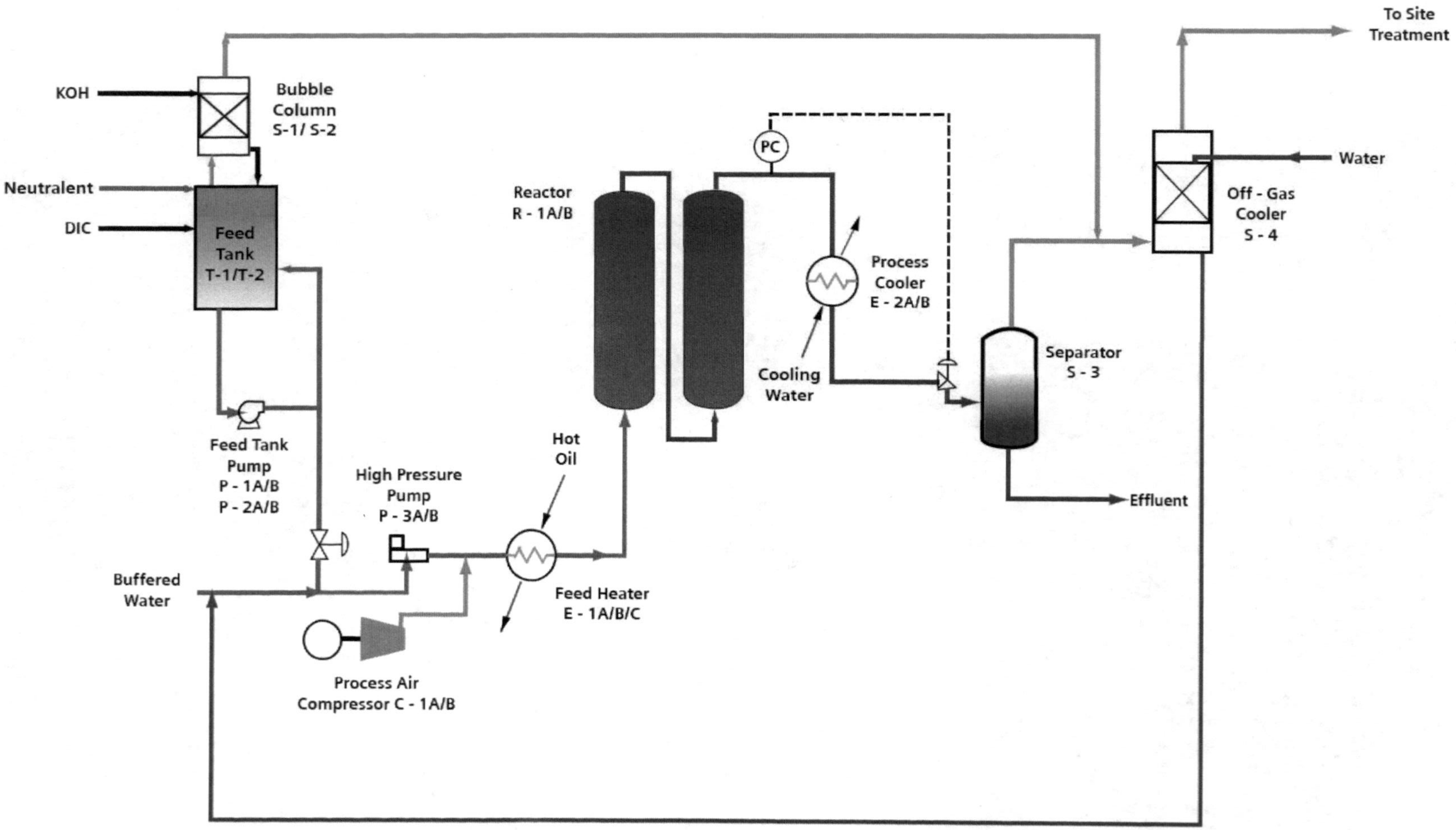

Figure 3. WAO System PFD

In operation, the neutralent is pumped from the storage tanks into one of two feed/mix tanks. Each tank has 7600 gallon capacity and is lined with a PTFE coating to withstand the fluoride in the feed. Potassium hydroxide and potassium bicarbonate/carbonate are added to the tank for pH control and as an oxidation enhancer. Dilution water is added to the feed for corrosion or oxygen demand control. After a prescribed mixing period, a high pressure pump is then used to feed the WAO system. The pumps and the system are rated to process up to 26.5 gpm.

Compressed air is added to the feed after the pump. The air compressors are rated to deliver around 850 lb air/ hr at 2500 psig. The system has redundant compressors and pumps so as to maintain maximum system uptime.

The air/neutralent fluid passes to the feed heater. The process tubes in the heater are made from nickel 200, to withstand the corrosive environment. A hot oil system is used as the heat source. The hot oil warms the feed to near the 300 °C reaction temperature.

The system has two reactor vessels in series, and the hot feed enters the base of the first reactor. Each reactor is a cylindrical vessel oriented vertically. The reactors are fabricated from steel and lined with nickel 200, to withstand the corrosive environment. The oxidation reaction releases sufficient heat to raise the temperature to the 300 °C operating temperature. Gas bubbles rise through the reactor and oxygen is transferred from the air bubbles into the liquid phase.

The oxidized material passes from the top of the second reactor through a process cooler. The coolers are similar to the heaters, with the exception that cooling water passes through the shells.

The cooled two-phase fluid passes through a pressure control valve and into a gas/liquid separator. The off-gas then passes to a water scrubbing gas cooler. The water from this water scrubber is recycled within the plant as dilution water for the feed. The off-gas passes through a thermal oxidizer before being vented. The oxidized liquid passes from the gas/liquid separator through an after cooler. The oxidized liquid passes to a deep well.

System Performance

Most of the feed materials oxidize very rapidly in the WAO system. MPA (methylphosphonic acid)is one of the reaction intermediate products and this material is the most recalcitrant carbon-phosphorus bond containing compound. During the 2001-2003 development work, it was found that essentially all residual C-P bonds were in the form of MPA. Therefore MPA analysis of the effluent was used as the measuring point for determining process effectiveness. Success criterion was established at <1000 ppm MPA in the effluent to meet U. S. Army and treaty obligations.

The QL neutralent was processed first, and by April 2007 all NSCMP QL neutralent was destroyed. Figure 4 shows a photograph of raw QL neutralent, the WAO feed after dilution and pH adjustment, and the oxidized WAO effluent. In this matrix the analytical detection limit of MPA was 50 ppm. No MPA was detected in the oxidized effluent, indicating that the MPA concentration was below 50 ppm. The treatment successfully met the <1000 ppm MPA success criteria.

After the QL neutralent campaign was complete, the WAO system was reconfigured to process the DF neutralent feed. This material had much lower chemical oxygen demand (COD) content, and so the air compressors were modified to deliver a lower flow rate of air. By November 2007, all of the DF neutralent was destroyed. Samples of the oxidized DF neutralent have been analyzed. The analytical sensitivity in this matrix is finer than in the QL neutralent. In this case, 24 ppm MPA was detected in the oxidized effluent. The treatment successfully met the <1000 ppm MPA success criteria.

System Reliability

Overall the process performance and reliability have been good and the project schedule is being met. There have been some mechanical maintenance items of note. In particular, the process air compressors experienced several mechanical failures and required maintenance under warranty from the compressor manufacturer.

The welded seams in the cladding in one of the two reactors leaked in multiple places. The leaks were detected through a weep-hole network that allows early detection of defects – before serious corrosion to the structural steel of the pressure vessel occurred. The leaks were due to manufacturing defects with a portion of the welds. There were no injuries, environmental release violations, or unacceptable process schedule delays resulting from the leaks. The defective welds were field repaired upon discovery. Once the welds were repaired in the initial months of the DF neutralent campaign, the unit operated at 93 percent availability enabling the project to be completed ahead of schedule.

Figure 4. QL neutralent, WAO feed prior to oxidation, WAO effluent.

Summary and Conclusion

The NSCMP and Shaw Environmental evaluated and selected WAO as the treatment technology of choice for the destruction of neutralized binary chemical warfare materials. A WAO system was installed in Houston, Texas and was operated from early 2007 through November of 2007 for the treatment of these materials. All NSCMP QL and DF neutralents were successfully treated and disposed of at the site. At the conclusion of this campaign, the WAO system was decommissioned, since the entire stockpile of these chemical warfare materials has been destroyed.

References:

Maugans, C.; Bettinger, J.; Adams, B.;Doyle, E. "The use of wet air oxidation to destroy neutralized chemical warfare material", CHISA conference, Prague, the Czech Republic, August 22-26, 2004.

Wet Oxidation as Applied to Powdered Activated Carbon Recovery in Wastewater Treatment

John A. Meidl, Siemens Water Technologies Corp., Rothschild, WI

ABSTRACT

Wet oxidation has long been used to destroy organics and sulfidic material in industrial wastes, and to treat sludges from both industrial and municipal applications. The process has also been shown to be successful in the recovery of powdered activated carbon (PAC) from wastewater applications. Through the years, powdered activated carbon recovery has been associated with thermal techniques such as fluid bed furnace or multiple hearth furnace technologies, in which the complicated process of dewatering and drying of the spent carbon prior to processing is required. With wet oxidation, however, all material remains in a slurry. And at the right operating conditions, sorbed and associated organics are destroyed, PAC is recovered for reuse, and autothermal regeneration occurs, making the process a non-fuel user.

The technical paper will discuss the influence of wet oxidation on powdered activated carbon characteristics, and how wet oxidation is applied to successfully recover the powdered carbon when it is used to treat wastewaters. Case studies will be included to verify wet oxidation's suitability for such applications.

KEY WORDS

Wet Air Oxidation, Wet Air Regeneration, wet oxidation, powdered activated carbon, PACT, carbon regeneration, wastewater treatment, Zimpro system

INTRODUCTION

Over the past couple of decades, wet air oxidation (WAO) has been refined and improved such that it has been found to be a useful tool in the treatment of concentrated industrial wastes, and in our immediate focus, for the recovery of powdered activated carbon (PAC) from wastewater applications. Challenges of the past have led WAO into the forefront for applications in the refinery and petrochemical markets for direct oxidation of wastes (example: spent caustic), and for the handling of sludges in the municipal and industrial markets (example: PACT® system). The success of WAO comes from its ability to destroy toxic/refractory organics and inorganics (sulfide) that conventional chemical and a biological processes cannot do. And it can do it without a particulate or NO_x/SO_x emissions problem. And without the use of much external fuel.

WHAT IS WET AIR OXIDATION?

Wet oxidation is the oxidation of soluble or suspended components in an aqueous environment using oxygen in the air as the oxidizing agent. The oxidation reactions occur at temperatures of 150°C to 320°C (275°F to 608°F), and at pressures from 10 to 220 Bar (150 to 3200 psi). The required operating temperature is determined by the treatment objectives.

The typical wet oxidation system is a continuous process using rotating equipment to raise the feed stream and air to the required operating pressure (see Figure 1). Heat exchangers are routinely employed to recover energy from the reactor effluent and use it to preheat the feed/air mixture entering the reactor. Auxiliary energy, usually steam, is necessary for startup and can provide trim heat if required. Since the oxidation reactions are exothermic, sufficient energy is usually released to allow the system to operate autothermally.

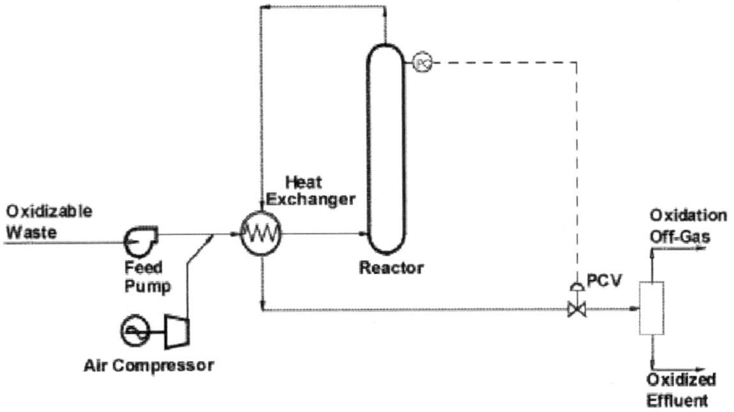

Figure 1 – Typical Process Flow Diagram

WAO will convert organic contaminates to CO_2, water, and biodegradable short chain organics, and oxidize inorganics which may involve any or all of the following reactions:

Organics + O_2	→	$CO_2 + H_2O$ + RCOOH*
Sulfur Species+ O_2	→	SO_4^{-2}
Organic Cl + O_2	→	$Cl^{-1} + CO_2$ + RCOOH*
Organic N + O_2	→	$NH_3 + CO_2$ + RCOOH*
Phosphorus + O_2	→	PO_4^{-3}

*short chain organic acids such as acetic acid make up the major fraction of residual organic compounds

WAO can be used to treat difficult process or wastewater streams to satisfy objectives such as:

- Gross reduction in COD
- Destruction of specific compounds
- Elimination of toxicity or reactivity
- Pretreat to produce readily biodegradable organics
- Process materials for recycle/recovery

The last objective brings us to the primary topic, Wet Air Regeneration.

WHAT IS THE WET AIR REGENERATION SYSTEM?

Wet Air Regeneration (WAR) is an aqueous phase oxidation process occurring at elevated temperatures and pressures that destroys excess biological solids, oxidizes carbon adsorbed organics, and 'reactivates' spent carbon for reuse within the PACT® system. It is able to do this by simply processing a gravity thickened spent carbon slurry.

When the destruction of various constituents is examined under wet air oxidation conditions, Figure 2, type of material and WAO temperature influence outcome. It is interesting to note that phenol, for example, is completely destroyed at about 220°C, but the COD associated with it is not. As explained previously, the compound is mainly oxidized to CO_2 and water, but some low molecular weight compounds (acetic acid) remain that contribute to the COD that remains.

Furthermore, examination of the graph shows that PAC is fairly resistant to WAO until WAO temperature exceeds 260°C. Since we know that organics associated with PACs from biological processes will oxidize before the PAC, the trick to successfully regenerating/recovering the PAC is to operate at a condition that maximizes destruction of sorbed and associated (biomass) organics while minimizing the oxidation of the PAC. Figure 3 shows how this is accomplished.

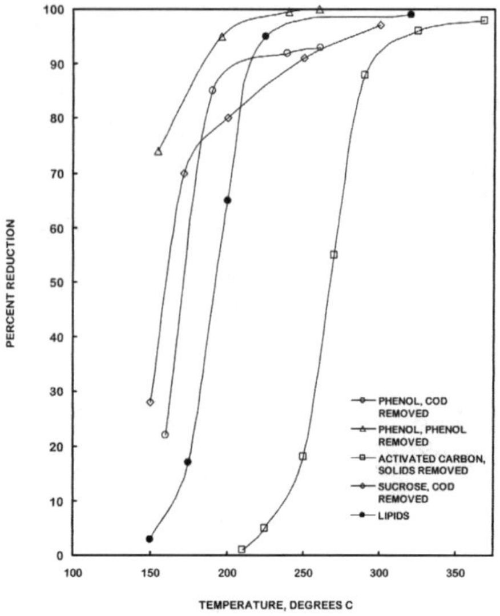

Figure 2 – Wet Air Oxidation Selectivity

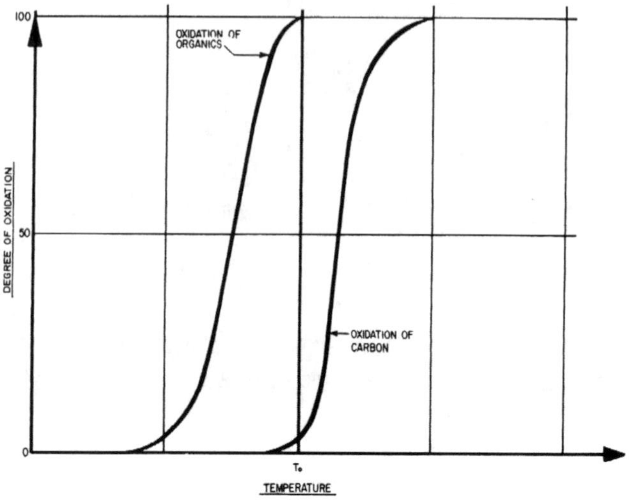

Figure 3 – Regeneration of Activated Carbon by Wet Air Oxidation

WET AIR REGENERATION INFLUENCE ON PACs

Though the oxidation/recovery of PAC from WAO has been discussed, the effects of WAO on the PAC's physical and chemical characteristics need to be examined.

Physical Characteristics

When PAC is wet oxidized, the particle size distribution is nearly identical except for a small increase in small particle diameters.

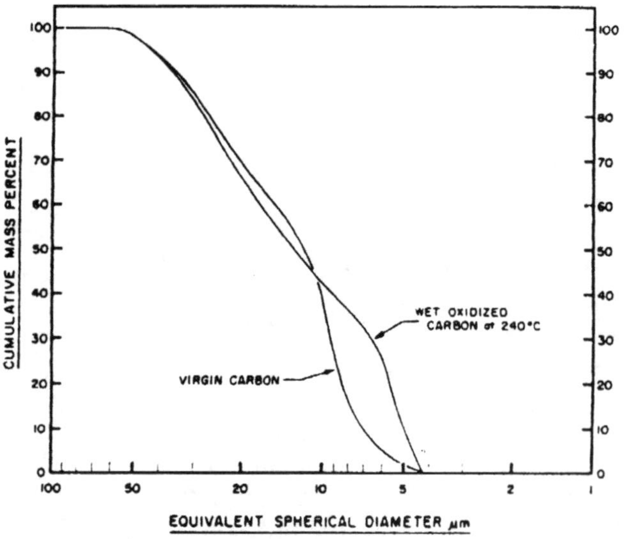

Figure 4 – Particle Size Distribution of a Wastewater Grade Carbon

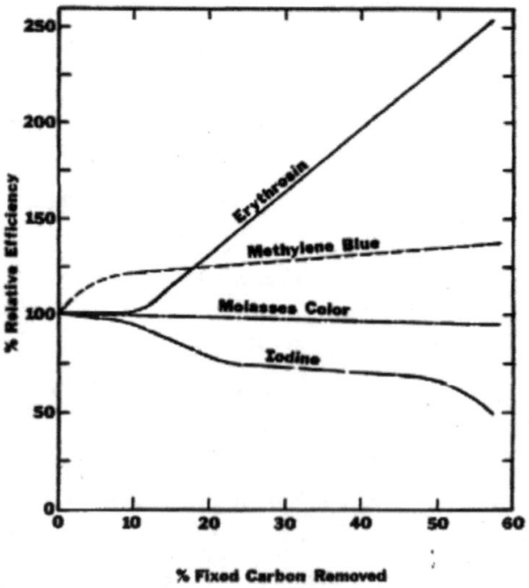

Figure 5 – Effects of WAO on Virgin Carbon

The average equivalent spherical diameter is 12 microns in the virgin carbon and 13 microns in the regenerated carbon. These results, at a regeneration temperature of 240°C (464°F), represent a severe oxidation condition for powdered activated carbon alone. Particle size distributions may vary from that shown in Figure 4 above at different regeneration conditions and different carbons.

Regenerated virgin carbons will also have a surface area and pore volume equal to or greater than virgin carbon in all pore diameter regions measured (see Table 1). In all pore diameter regions, the regenerated powdered activated carbon pore volume is greater than the starting carbon and increases with higher regeneration temperatures. The real density of the powdered activated carbon, as measured in helium, is slightly increased by oxidation.

Table 1 – Physical Properties of Wet Air Regenerated Powdered Activated Carbon

	Virgin Carbon	Wet Air Regenerated Carbon		
Temperature, °C (°F)		200°C (392°F)	215°C (419°F)	230°C (446°F)
Specific Areas of Pores, m^2/g				
PD < 37 Å *	363	395	372	367
PD 37 – 600 Å	84	89	161	112
PD > 600 Å	6	6	7	6
Total	453	490	540	485
Specific Pore Volume, cm^3/g				
PD < 600 Å	0.191	0.244	0.308	0.258
PD > 600 Å	0.939	1.016	1.109	1.214
Total	1.130	1.260	1.417	1.472
Real Density, g/cm^3	1.93	--	--	1.95

* PD = Pore diameter
A = Angstrom (10^{-10} meters)

To better define the smallest pore diameter region of powdered activated carbon, chemical adsorbate tests for iodine, methylene blue, erythrosin and molasses color were selected. Repeated wet air oxidations of commercial PACs were done to study the impact of repeated regenerations of the adsorptivities of those compounds.

As wet oxidations proceed, the relative capacities change as shown in Figure 5. Note that there are significant increases in the methylene blue and erythrosin adsorptivities exceeding the adsorptivity of the starting carbon. The relative capacities for molasses remain quite stable, whereas there is loss of iodine chemical adsorptivity during repeated wet air regenerations.

Thus, one can conclude that surface area in the less than 10 Å pore region decreases, in the 10-30 Å pore region increases, and the > 30 Å pore region changes very little.

The wet air regeneration of a virgin carbon produces a carbon similar to or superior to the starting carbon. The pore volumes, surface areas and chemical adsorptivities are increased by wet air regeneration and continue to increase with higher temperatures.

Chemical Characteristics

When PACs are wet oxidized, oxygen utilization exceeds solids removal at all temperatures. Why? Oxygen containing groups form part of the insoluble PAC macromolecule. Table 2 shows the change in composition, forming a new type of structure, with some replacement of carbon by oxygen. Though not as pronounced, the same is true for hydrogen and nitrogen that are found in the air supplied to the WAR system.

Table 2 – Composition of PAC Solids

	Virgin Carbon	Wet Air Regenerated Carbon		
Temperature, °C (°F)		200°C (392°F)	215°C (419°F)	230° (446°F)
Inorganic % WT/WT	24.9	22.6	23.5	24.4
Atomic Ratios in Organic Fraction				
C	1	1	1	1
H	0.19	0.23	0.26	0.29
N	0.008	0.009	0.010	0.011
O	0.07	0.13	0.16	0.20

Because of the 'adds' of oxygen, nitrogen and hydrogen to the carbon macromolecule, this new structure now becomes less prone to oxidation within the WAR process.

The above results are all academic in nature since the purpose of the WAR process is to recover the PAC for reuse in the PACT® wastewater treatment system and yield a regenerated product that produces a high quality effluent out of the treatment system in which the recovered PAC is placed.

Performance of the WAR system may differ from the previous results since the quantity and characteristics of the extraneous solids in the wastewater to a PACT® system affect the mixed liquor and spent carbon slurry characteristics. Due to the presence of suspended inerts, the overall particle size distribution of the carbon slurry (powdered activated carbon + inerts) may vary as may the chemical adsorptive characteristics. Slightly higher regeneration temperatures may also be used for PACT® system spent carbon, again, depending on spent carbon slurry characteristics.

Wet air regeneration of PACT® system spent carbon / biomass slurries results in some loss of surface area due to accumulation of extraneous solids. However, effective PACT® system treatment is maintained. Pore volumes are increased during wet air regeneration, exceeding the staring carbon in all pore regions. The real density of carbon increases due to the presence of the extraneous solids, an increase in organically bound oxygen, and adsorbed organo-metallic compounds resulting from repeated adsorption and regeneration.

The recovery of adsorptivity in regeneration of PACT® system spent carbon, from numerous pilot and full scale treatment applications, is shown in Table 3 for temperatures from 200 to 250°C (390° to 480°F). At about 250°C, the molasses color adsorptivity is generally comparable to the virgin parent carbon. The loss of some adsorptivity in the smallest pore regions (iodine chemical adsorptivity) is noted but is of little concern in the PACT® system treatment since the lowest molecular weight wastewater organics are readily biotreated. The molasses adsorptivity is considered indicative of the adsorption of organics of concern in waste treatment.

Table 3 – Temperature Effect on Chemical Adsorptivities of Regenerated Carbon from PACT® System Sludges

	Regenerated Carbon		
Temperature, °C (°F)	200°C (392°F)	230°C (446°F)	250°C (482°F)
Adsorbate Relative Efficiencies, % of Starting Carbon			
Iodine	20-40	45-60	60-80
Methylene Blue	30-60	60-90	85-135
Erythrosin	30-60	50-130	80-110
Molasses	50-85	70-110	100-150

The most important measure of regeneration system performance is the performance of the PACT® system which it supports and will be discussed under case studies.

Ash Characteristics

Removal of inorganic solids (inerts) from the WAR system is done to control inerts buildup in the PACT® treatment system. Inerts are comprised of inorganics (calcium, magnesium, etc., originally associated with the raw wastewater), inorganic precipitates (sulfates, phosphates, carbonates) formed during wet air regeneration, inorganic biomass residuals, and some powdered carbon.

The quantity of ash blowdown varies and depends mainly on the wastewater characteristics to the treatment system. The ash (inerts) removed ranges from a coarse to a very fine sand that can be effectively removed and dewatered.

Important is the leachability of various heavy metals from ash blowdown solids. Table 4 shows metal concentrations in an ash blowdown sample collected from a full scale PACT system treating a combined domestic/metal plating and finishing waste. Despite high metals concentrations in the Wet Air Regeneration blowdown, the soluble content of these metals is quite low. And when the blowdown is subjected to testing to determine its acceptability for landfilling (denoted as RCRA Extract of Blowdown), its concentration is lower than RCRA limits meaning that the blowdown can be safely disposed in a conventional landfill.

Table 4 – Metals Analysis of Wet Air Regeneration Reactor Blowdown*

Metal	Total Concentration, mg/l	Soluble Concentration, mg/l	RCRA Extract of Blowdown Sample Total Concentration, mg/l	RCRA Limits Total Concentration, mg/l
Arsenic	0.130	< 0.004	0.024	5.0
Silver	3.6	< 0.10	0.02	5.0
Barium	308	0.03	0.42	100
Aluminum	51,100	0.40	3.82	--
Calcium	50,900	5.52	14.5	--
Copper	680	0.12	0.48	--
Chromium	206	0.04	0.03	5.0
Cadmium	15	0.02	0.01	1.0
Mercury	0.12	< 0.008	< 0.002	0.2
Lead	918	0.04	0.39	5.0
Magnesium	10,565	19	6.99	--
Selenium	0.081	< 0.004	0.0048	1.0
*From PACT treatment of a domestic/metal plating and finishing waste.				

Off-gas Characteristics

The Wet Air Regeneration system off-gas (spent air) is collected and typically treated in a single stage liquid scrubber followed by transfer to the PACT system's aeration blower inlet for subsequent aeration mixed liquor scrubbing.

Predicted Wet Air Regeneration off-gas characteristics for the PACT® system are shown in Table 5. The scrubbed gas characteristics may be expected to vary depending on regeneration conditions and wastewater inlet constituents.

Table 5 – Wet Air Regeneration Off-gas Characteristics (from processing PACT® System Sludge)

Sample Point	Concentration in Percent Volume				Total Hydrocarbons, % (minus CH_4 as CH_4)
	CO_2, %	O_2, %	N_2, %	CO, %	
Spent air before treatment	14 - 16	1.0 - 3.0	78 - 80	0.5 - 1.5	0.1 - 0.4
After single stage water scrubbing	12 - 14	1.0 - 3.0	80 - 82	0.05 - 1.5	0.05 - 0.3
After aeration tank scrubbing	2 - 5	16 - 20	78 - 80	0.05 - 1.5	0.005 - 0.05

CASE STUDIES

PAC regeneration support for PACT® systems is well known. A couple case studies are reported herein that verify WAR's suitability for the efficient recovery of powdered carbon's from PACT wastewater applications.

Vernon, CT – Municipal

In 1978, the Town of Vernon, CT upgraded its wastewater treatment plant from a trickling filter system to a 24,000 CMD (6.5 MGD) PACT/WAR system in order to meet stringent Hockanum River BOD, SS, and color standards. At the time of the conversion, Vernon wastewater contained mainly industrial wastes from textile dye houses, and minor amounts of wastes from electrical control and printed circuit board manufacturing and metal anodizing. Wastes to the treatment plant were highly variable and highly colored. The dyeing operations were primarily engaged in nylon fabric dyeing and finishing with smaller emphasis on other synthetic fabrics including polyester, polyester-cotton blends, acetate, and cellulosic fabrics. The principle acid dyestuffs included orange, red, blue, navy and black dyes. Dispersed and direct dyestuffs were also used.

The secondary/tertiary treatment system flow diagram, Figure 6, was initially operated as a conventional activated sludge (AS) system prior to the WAR system and the PACT system being placed on-line. All AS solids were handled and disposed with the primary solids. Once the WAR system construction was completed, conversion from AS to PACT occurred.

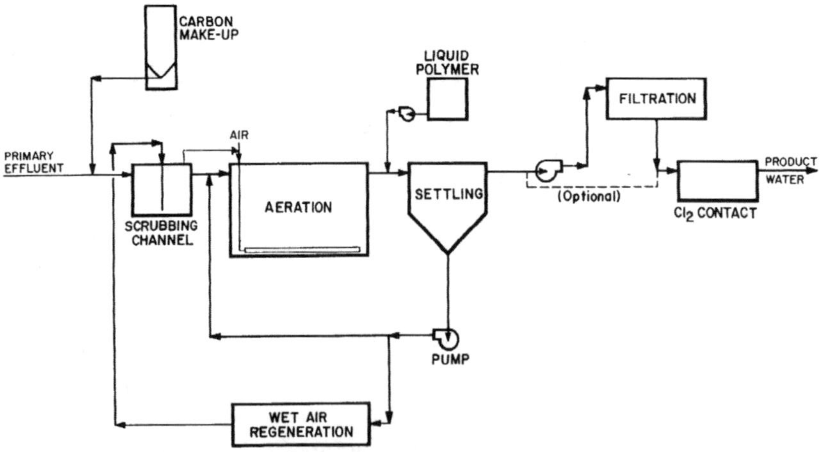

Figure 6 – PACT/WAR System – Vernon, CT

Operating information and performance of the system when operated as an AS system and a PACT/WAR system are shown in Tables 6 and 7. Though operation as AS was stable, it did not remove sufficient color.

Table 6 – Wastewater Treatment Plant Operation – Vernon, CT

	As Activated Sludge	As PACT
Hydraulic Detention Time (HDT), hrs	6+	3
Solids Residence Time (SRT), days	3	5
Mixed Liquor Suspended Solids (MLSS), mg/L	3,000	12,000
Return Sludge Concentration, %	< 1	3 – 5

Table 7 – Wastewater Treatment Plant Performance – Vernon, CT

	Raw Wastewater	Activated Sludge Effluent	PACT Effluent
BOD_5, mg/L	150 – 200	33	< 5
COD, mg/L	700 – 900	240	< 80
SS, mg/L	300 – 500	39	< 20
Turbidity, NTU	100 – 200	20	< 10
Color, APHA	100 – 600	> 100	< 50

Conversion of the system to a PACT/WAR system occurred in early 1979. The PACT system was able to successfully treat the industrial/municipal wastewater and remove the problematic dyes. The Hockunum River, which once was no more than a colored channel of partially treated wastewater in which no life could exist downstream of the WWTP, changed to fishable river after the PACT system went on-line.

During 1980-1985, the WAR system was closely monitored. During this time, PACT system sludge sent to WAR was 47.9 x 10^6 lbs of which 11.5 x 10^6 lbs was carbon. Virgin carbon makeup over this period was 0.80 x 10^6 lbs. Carbon oxidation losses within the WAR system were < 5%. Overall carbon makeup losses for the entire system were 8% - the difference being solids that needed to be wasted from the WAR unit in order to control ash buildup in the PACT system.

The PACT/WAR system was expanded/upgraded in 1998 to treat 27,000 CMD (7.1 MGD) in order to meet new nitrogen standards imposed by the Connecticut DEP. HDT was increased to 6-7 hours to enable a SRT of ~20 days, allowing for full nitrification and partial denitrification. Recent data from the plant, shown below in Table 8, attests to the viability of both the PACT system and the WAR system.

Table 8 – PACT/WAR System Performance

	Effluent
BOD	< 5
COD	< 50
SS	< 10
Turbidity	< 10
Color	< 50
NH_3-N (Ammonia)	< 0.1

<u>Renda Industrial Park, TaSheh, Taiwan</u>

In 1995, the Renda Industrial Park (RIP) was required to expand and upgrade its existing 10,000 CMD (2.6 MGD) primary + activated sludge system to 23,000 CMD (6 MGD) in order to serve an expanding client base, and at the same time to meet a new < 100 mg/l effluent COD standard. However, the extended air activated sludge unit, even with chemically assisted precipitation, was only able to produce an effluent of 150 – 250 mg/l COD.

Because a majority of its wastes come from chemical and petrochemical companies, the waste has an extremely low BOD/COD ratio (less than 0.1). RIP pilot tested a number of technologies and chose the PACT/WAR system for implementation.

Powdered carbon (PAC) adsorption testing and PACT system testing with and without WAR were conducted. Initial testing used three different PACs. One was chosen to do extensive on-site pilot testing of the PACT/WAR system. A summary of the results is shown in Table 9.

Table 9 – Powdered Activated Carbon Testing – Renda Industrial Park, Taiwan

		Effluent COD, mg/L Carbon Type		
Type of Test		**1**	**2**	**3**
Adsorption (Inlet COD = 850 mg/L)		273	292	304
PACT® System (Inlet COD = 850 mg/L):	Condition 1	80	80	103
	Condition 2	61	69	92
On-site PACT/WAR System (Inlet COD = 600 – 900 mg/L)		< 80	--	--

The PACT/WAR system was chosen for the RIP expansion and upgrade for the following reasons:

- Meets < 100 mg/L effluent COD requirement
- Able to convert existing WWTP with minimal expansion
- Small footprint
- The primary treatment system did not require expansion
- Lower O & M than all other options
- Minimal amount of sludge to disposal
- Odor control provided for biological and sludge systems at minimal cost.

Even though flows have increased since the activated sludge days, there has been a significant reduction in solids to disposal as shown in Table 10.

Table 10 – RIP Sludge to Disposal, KG/d Wet Solids

	1995[1]	Current[2]		
		2004	**2005**	**2006**
Primary + Activated Sludge with Belt Press Dewatering	13,000			
Primary + PACT/WAR with Sludge Drying Bed		790	614	692

[1] secondary treatment
[2] tertiary treatment

The following are relevant to the WAR system:

- Gravity thickened spent carbon slurry fed to the WAR system is in the range of 10 ~ 15% solids.
- WAR system operates autothermally, requiring no outside source of fuel.
- Ash is separated from the regenerated powdered carbon via a small physical-chemical process.
- > 90% of the PAC is recovered for reuse.
- A slurry of ash is the only material wasted to a drying bed; dried solids are landfilled.

PACT/WAR performance has remained steady since it first began operation in 1996. Data from the last few years is shown in Table 11 below.

Table 11 – Treatment Results Powdered Carbon/WAR System, Renda Industrial Park

	Equalization Tank Effluent (PACT System Influent)			Effluent Discharge		
	COD mg/l	BOD mg/l	SS mg/l	COD mg/l	BOD mg/l	SS mg/l
Averages	236	26	52	75	2	8

It is interesting to note that the original plant was only able to remove 50% of the COD applied when operated at about the same aeration detention time the PACT system currently operates at, even though the BOD/COD ratio during the AS days was 0.5.

The current total operating cost for the entire wastewater plant (including all labor, chemicals, electrical power, fuel, domestic water, maintenance, equipment replacement, benefits, sludge disposal, etc. is approximately $0.94 USD/1000 gal.

ACKNOWLEDGEMENTS

Thank you to the following personnel for their input into the case studies presented herein.

Arnie Bevins, WWTP, Superintendent and Assistant Director, Vernon Water Pollution Department, Vernon, CT

K.L. Chou, President, and Howard Juang, Sr. Engineer, Jinn Wei Engineering Company, Kaohsiung, Taiwan

71d Comparison of Treatment Efficiencies of Various Advanced Oxidation Processes for Dye Removal from Wastewater

Meng-Wen Chang, Pei-Jen Lu, Tai-Shang Chen, and **Jia-Ming Chern**. Department of Chemical Engineering, Tatung University, 40 Sec. 3, Chung-Shan N. Rd., Taipei, Taiwan

Textile wastewaters are usually treated by biological processes to removed most organic compounds (BOD), followed by advanced treatments to remove remaining dye to meet strict COD effluent standards. Although some advanced treatment such as activated carbon adsorption is quite effective to remove dye, dye-saturated activated carbon is not easy to be regenerated and is classified as hazardous solid waste. Advanced oxidation processes (AOPs) using powerful hydroxyl radicals to non-selectively decompose organic compounds in wastewaters have received great attention in industry. In this study, various AOPs, namely UV/H_2O_2, UV/TiO_2 (suspended and immobilized), Fenton's reagent, and ozone were employed to decompose pink acid dye in simulated wastewaters. The experiments were carried out in a batch reactor with varying operating conditions and the dye concentration was continuously monitored to compare the performance of different AOPs. Among the AOPs tested, the ozone system has the highest dye removal efficiency. The experimental results also show that the dye decomposition rate increases with increasing initial dye concentration. Increasing the reaction temperature can increase the treatment efficiencies in UV/TiO_2 and UV/H_2O_2 systems, but decrease the treatment efficiencies in Fenton's reagent and ozone system. Higher solution pH favors the treatment efficiency in the ozone system but does not favor the UV/TiO_2, UV/H_2O_2 and Fenton treatment systems. The results of this study can be used to choose a proper AOP to treat dye-containing wastewaters.

Design of a static mixer for short, fat, laminar flow reactors

Matthew Armstrong and E. Bruce Nauman

Department of Chemistry and Life Science
United States military Academy
400J Bartlett Hall
West Point, NY 10996
Phone: (845) 938-2740
Email: matthew.armstrong@usma.army.mil

The Isermann Department of Chemical and Biological Engineering
Rensselaer Polytechnic Institute
Troy, NY 12180

Abstract

The focus is on sterilizations and similar reactions where all portions of the fluid must experience some minimum residence time. The minimum residence time corresponds to the first appearance time measured by tracer experiments. Short, fat reactors are desired for economic reasons, but such reactors can have a first appearance time less than expected for fully developed laminar flow. This is a classic problem in entrance length flows. Static mixers can increase the minimum residence time, but conventional, proprietary designs are ill suited for the purpose. CFD was used to test a simple design that uses axisymmetric baffles to slow the fastest moving fluid elements and thus to increase the first appearance time.

Keywords

Residence Time, First Appearance Time, Sterilization, Retrofit, Reynolds Number

Introduction

Effective sterilization requires a minimum reaction time. This is easy to provide in a batch reactor, but is more difficult in a flow system. The minimum residence time can in principle be found from hydrodynamic calculations or tracer experiments. For fully developed laminar flow in a tube, the first appearance time is half the mean residence time, $t_{first} = \bar{t}/2 = L/\bar{u}$. There are a variety of ways that t_{first} can be increased, the simplest being to increase $\bar{t}$, but better alternatives may exist. Define the normalized or dimensionless first appearance time as $\tau_{first} = t_{first}/\bar{t}$. Then the goal is to increase τ_{first} above its nominal value of 0.5 while keeping a simple, streamlined design. Helically coiled tubes have higher values for τ_{first}, approaching 0.613 in the limit of a long tube (Nauman & Buffham). If the axis of the helix is systematically changed in direction, $\tau_{first} > 0.85$ is possible (Nigam). Commercial static mixers also increase τ_{first}, but none of these approaches is well suited for the short, fat reactors that are the cheapest to build. The problem is compounded for short, fat reactors because entrance effects can cause $\tau_{first} < 0.5$. To counteract this, a design consisting of axisymmetric baffle designs was developed.

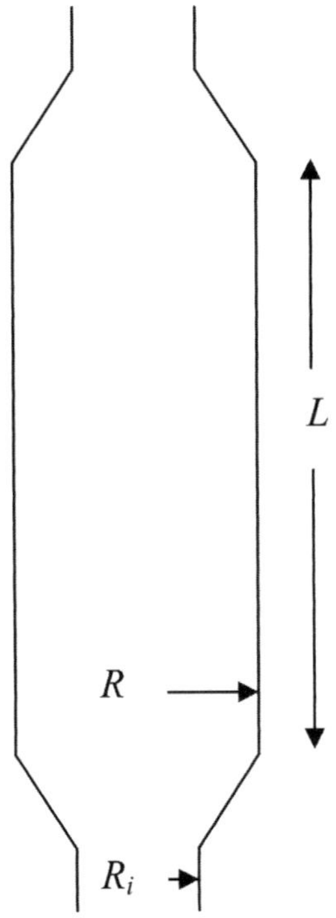

Parameters to be varied:

$L/R = 2, 4, 8, 16, 32$

$R_i/R = 1/16, 1/8, 1/4, 1/2, 1$

With and without baffles

Cone Angle = 45°

Required Result

$\tau_{first} = t_{first} / \bar{t}$

Figure 1: Dimensionless values for treatment tank

Figure 1 shows the design of what we have termed a short, fat reactor. The straight run portion has length L and radius R. The inlet diameter is R_i. The case of a cone angle of 45° was considered. The reactor is completely characterized by one absolute dimension, say R, and two dimensionless ratios, L/R and R_i/R. The volume of the reactor for the 45°cone inlet and outlet is

$$V = \pi R^3 \left[\frac{2}{3} \left(1 - \frac{R_i^3}{R^3} \right) + \frac{L}{R} \right]$$

Parameters to be varied:

R_b/R = 2/10, 4/10, 2/3, 8/10, 9/10

L_{Cone}/L_{Tank}=4/25, 2/10, 9/50, ¼

Number of Baffles = 1, 2, 3 spaced evenly along centerline

Figure 2: Dimensionless values of baffle

Figure 2 shows the baffle design. An additional parameter, R_b/R, is needed to characterize the baffle. The height of the baffle is kept constant. The volume of the baffle is:

$$V_{baffle} = \frac{2}{3}\pi R^3 \left(\frac{R_b}{R}\right)^2 * \frac{L_{cone}}{R}$$

Also needed are the number and spacing of the baffles within the straight run portion. A CFD program, Fluent, was used to model flow within the reactor. The input parameters were chosen to give creeping laminar flow, **Re** < 2, even in a tube of diameter R_i. Numerical studies were constrained such that $.2 \le R_i / R \le 1$. The flow is axisymmetric so only a two-dimensional solution is needed. Single, double, and triple baffle cases were considered. The baffles were evenly distributes along the centerline. The numerical calculations gave values for τ_{first} where τ_{first} is the first appearance time normalized by the mean residence time, $\bar{t}$, in the reactor before the baffles were installed.

This is the retrofit case. Since the baffles take volume, the mean residence time in the retrofitted reactor will be lower than $\bar{t}$. Another possibility is to increase the length of the reactor when baffles are installed to give the same active volume and thus the same value for the mean residence time. This is the constant volume case.

The velocity profile at the reactor inlet is assumed to be flat, but it quickly reflects the wall boundary conditions in the conical entrance region. It becomes parabolic – at least for the case without baffles – in the straight portion of the reactor. Figure 3a shows the velocity profiles at four downstream locations for a reactor with $R_i / R = 0.33$.

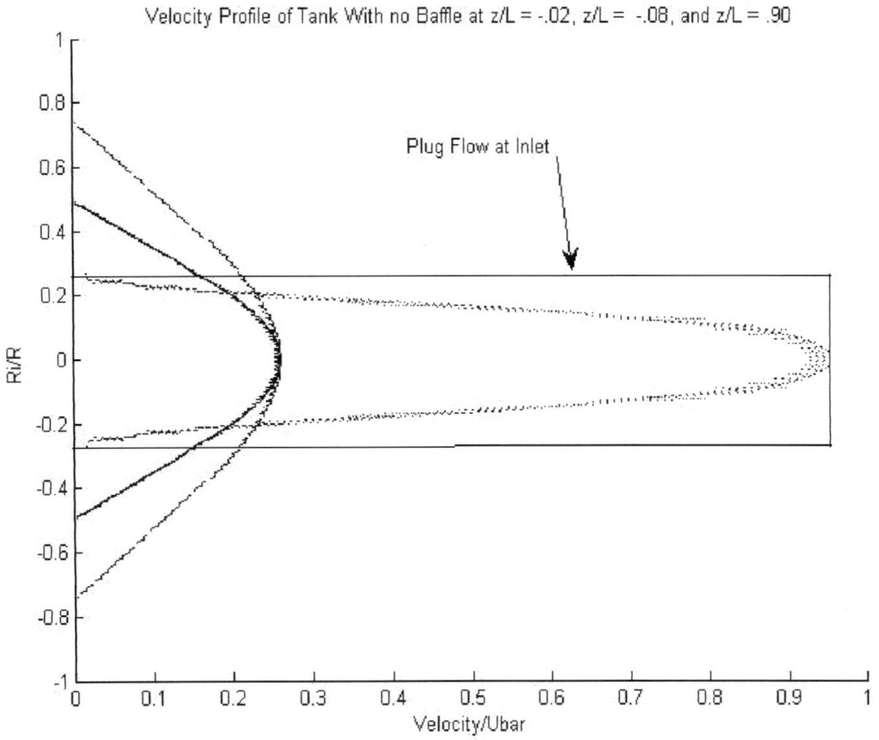

Figure 3a: Velocity Profile at 3 Locations in Reactor

Results and Discussion

First, a series of cases were run to illustrate the entrance length problem with the tank geometry shown in figure 3b. For a simple pipe, with fully developed laminar flow at the inlet and without a restricted aperture, the normalized first appearance time should be 0.5. This geometry causes an entrance length problem for two reasons: 1) the flow at the inlet is plug flow and takes a certain minimum distance to become fully developed inside the tank; and 2) the restricted inlet aperture of the inlet magnifies the problem. Figure 3b shows that to approach a normalized residence time of 0.5 for a 45° inlet cone angle and $R_i/R = 0.33$ would require an L/R ratio approaching 12 - 15. Such a reactor would be more expensive than a shorter, fatter design. By adding baffles along the centerline we can approach and surpass first appearance times of 0.5, without increasing the physical length of the treatment tank.

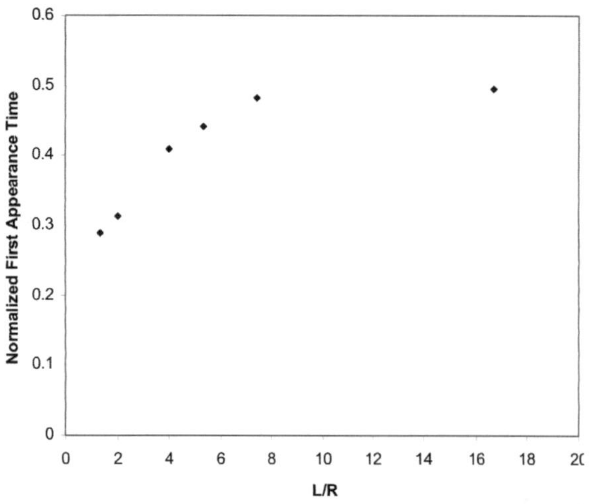

Figure 3b: Exploring Entrance Length Effects

The graphs shown below in are normalized first appearance time versus normalized radius. The radius values are normalized by the tank radius. The series of plots was created for L/R values of 2.66, 4, 7.46, 8 and 10, for different values of R_i/R. Plots of single, double, and in some cases triple baffle systems are overlaid. The general shape of the series of plots remains consistent. These plots for $R_i/R = 0.33$ are shown in Figures 4a, 4b, 4c, 4d and 4e.

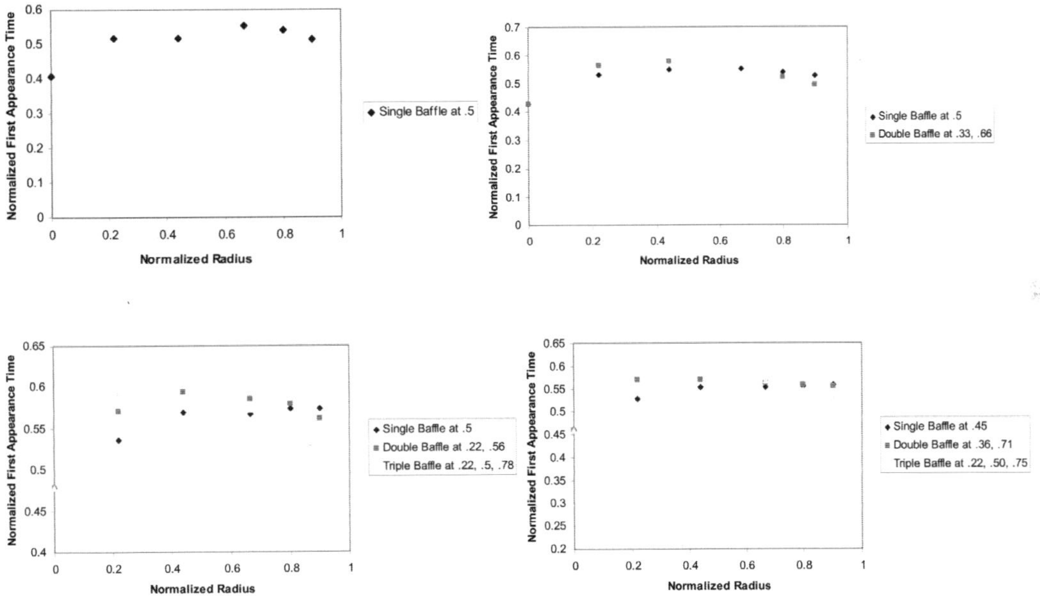

Figures 4a, 4b, 4c, 4d, 4e Normalized First Appearance Time vs. Normalized Radius for $R_i/R=0.33$ (All baffle distances are dimensionless)

The increasing then decreasing shape of the curve is due the fact that the baffle consume what would otherwise active reactor volume. These plots are for the retrofit case and are based on the original, open-tube value for $\bar{t}$, which was used to normalize first appearance times. Figure 5 shows the effects of 1) keeping L/R constant but varying actual length and actual tube radius, and 2) varying the tube length to keep a the active volume constant.

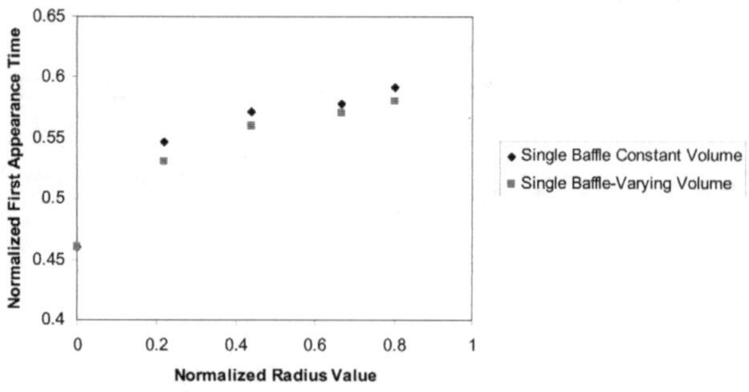

Figure 5 Comparison of First Appearance Times Keeping Active Volume Constant

Figure 6 shows the effect of baffle radius in a reactor with one baffle. In this plot L/R = 7.46, and the baffle is placed halfway down the treatment tank along the centerline

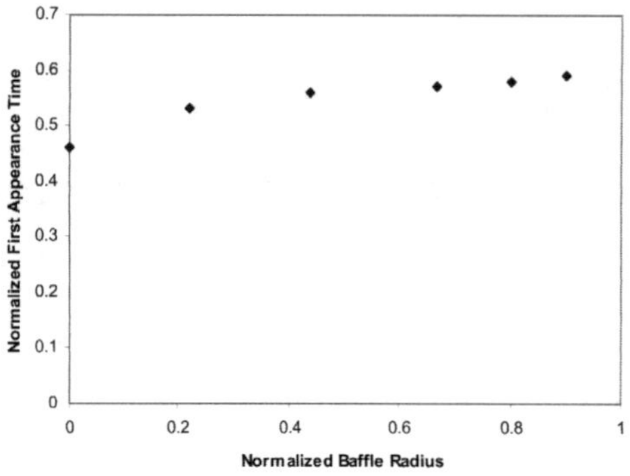

Figure 6 Effect of Varying Baffle (Radius) Size in a Single Baffle System

Conclusions

The goal of this research was to maximize the first appearance time of a laminar flow reactor system. Due to the original short and fat geometry, and its associated entrance length problem, depicted in Figure 3b, the fluid streams towards the middle of the tank sped through in less than first appearance times of 0.5. To maximize the first appearance time, axisymmetric baffles were added along the centerline of the tank.

As the number of baffles increased along the centerline, first appearance time was maximized. With respect to the results shown here the largest first appearance times were created with triple baffle systems. Over the ranges studies, the inlet aperture size had little effect on first appearance times and an increased baffle radius led to an increase in first appearance time. With respect to the single baffle and double configurations, physical placement of the first and second baffles within the waste water treatment tank did not effect first appearance time by more than 1%.

AIChE 2008 Annual Meeting, New Orleans, LA
April 8, 2008

Environmental Regulatory Hot Topics: New Source Review and Recent Judicial Developments

Robert E. Holden

(504) 556-5140 Direct
REHolden@Liskow.com

LISKOW&LEWIS

A Professional Law Corporation

New Orleans | Lafayette | Houston

What is New Source Review?

- Permitting program for new and modified major stationary sources
- Requirements vary depending upon air quality in area of proposed project
  - Prevention of Significant Deterioration (PSD) – attainment areas
  - Nonattainment NSR (NNSR)– nonattainment areas

Evolution of NSR

- 1972 – DC Federal District Court rules that EPA must ensure that air quality does not deteriorate in areas that meet NAAQS. *Sierra Club v. Ruckelshaus*, 344 F. Supp. 253
- 1974 – EPA promulgates first PSD rules
- 1977 – Congress amends CAA to add NSR
- 1978 – EPA rules implementing 1977 CAA amendments
- 1979 – *Alabama Power Co. v. Costle*
- 1980 – EPA revises NSR rules in response to *Alabama Power*

Evolution of NSR (cont'd)

- 1989 – *Puerto Rican Cement Co. v. EPA* (PSD enforcement case upholding actual-to-potential policy)
- 1990 – *Wisconsin Electric Power Co. v. Reilly ("Wepco")* (requiring actual-to-projected actual)
- 1990 – CAA Amendments
- 1992 – EPA amends NSR rules pursuant to *Wepco* opinion for EGUs
- 1996 – "NSR Reform" rules proposal
- 2002 – EPA promulgates NSR Reform Rules

Evolution of NSR (cont'd)

- 2005 – *New York v. EPA I* (NSR reform ruling upholding actual-to-potential, 10-year baseline, demand growth, and PALs; vacating Clean Units and Pollution Control Project exemptions)
- 2006 – *New York v. EPA II* (vacating ERP)
- 2007 – *Environmental Defense Fund v. Duke Energy*, 127 S.Ct. 1423 (reversing 4th Circuit and upholding actual-to-potential test)
- 2007 – *Massachusetts v. EPA*, 127 S.Ct. 1438 (GHGs are "pollutants" under CAA)

The Big Issues

- New sources vs. modifications
- What is a physical change or change in the method of operation?
- If an activity is a physical change or change in the method of operation, how do we calculate the emissions increase?

New York v. EPA I and 2002 NSR Reform

- 10-Year Look Back Period – *O.K.*
- Plantwide Applicability Limits– *O.K.*
- Actual-to-Projected Actual Test (with small exception) – *O.K.*
- Demand Growth Exclusion– *O.K.*

==============================

- Pollution Control Projects -- *No*
- Clean Unit Applicability Test --*No*

10-Year Baseline Emissions

- Any two-year period
- Allows for full economic cycle
- Allows different periods for different pollutants
- Requires correction for new emission standards
- How does this loosen NSR?

LISKOW&LEWIS

New Orleans | Lafayette | Houston

It Depends on What *"Increase"* Means – The Weather Analogy

- What is an "increase" in temperature?
- Night time to day time, day to day or season to season?
- "Increase" is vague enough to allow EPA to exercise discretion
- The baseline requires some technical judgment
- A new administration has the right to make different policy choices

The Demand Growth Exclusion

- Derives from the *Wepco* rules
- "Post change emissions calculation excludes emissions increases that an existing unit ***could have accommodated*** during the consecutive 24-month period used to establish the baseline actual emissions, and that are also ***unrelated*** to the particular project, including any increased utilization due to product demand growth."

LISKOW&LEWIS New Orleans | Lafayette | Houston

How Will the Demand Growth Exclusion be Applied?

- A product fad – O.K. (per preamble)
- If unrelated, does this allow the use of pre-project PTE?
- How close is this to a potential-to-potential rule?

Plantwide Applicability Limitations (PALs)

- 10-year permit
- 2-year baseline plus significance threshold
- Applies to whole plant
- Recordkeeping requirements
- 10 years necessary to lock in the benefits of the PAL process

New York v. EPA II and the ERP

- Longstanding "routine maintenance" exemption
- The Equipment Replacement Provision (ERP):
  - Identical replacements;
  - Up to 20% of capital costs of unit; and
  - No change to purpose of unit.
- Vacated by D.C. Circuit
- Consequently, no "bright line" test for what is "routine" maintenance

Narrow Test : "Routine" means "frequent for the unit"

- Like an oil change for your car.
- *United States v. Ohio Edison Company*, 276 F. Supp. 2d 829 (S.D. Ohio 2003) (upholding "routine" for the unit test)
- *United States v. Southern Indiana Gas and Electric Company*, 245 F. Supp. 2d 994 (S.D. Ind.2003) (upholding routine for the unit test)

Broad Test: "Routine" means "frequent for the Industry"

- Like installing a new transmission in your car.
- *United States v. Duke Energy Corp.*, 278 F. Supp. 2d 619 (M.D.N.C. 2003), *aff'd*, 4th Cir. (June 15, 2005) ("routine within the industry considering the facts")
- *U.S. v. Alabama Power*, No. 01-152-VEH (D. Ala. June 3, 2005) (routine within the industry)

Environmental Defense v. Duke Energy

127 S.Ct. 1423 **(April 2, 2007)**

- History
- Issues
- What the Supreme Court Case Did Not Address

History

- District Court opinion on cross motions for summary judgment
- U.S. stipulated no hourly increase in emissions
- Appeal to 4th Circuit
- 4th Circuit Opinion

Fun Points from Supreme Court Opinion

- Fourth Circuit's construction of PSD regulations was "implicit invalidation"
- Challenge to regulations under CAA § 307(b)?
- Fourth Circuit opinion vacated, case remanded

The Supreme Court Case Did Not Resolve

- Routine maintenance!
- Statute of Limitations

Massachusetts v. EPA

- Mobile source decision
- Greenhouse gases, including CO_2, are "pollutants" under CAA
- Waiting for the NSR shoe to drop!

PSD Applicability

- "Major Sources" with potential to emit 250 tons per year or 100 tons per year from listed source categories of any ***"regulated NSR pollutant"***
- "Major modification" – Any p*hysical change* or change in method of operation that results in *significant net emissions increase* over threshold for any ***"regulated NSR pollutant"***

"Regulated NSR Pollutants"

- Excludes Section 112 HAPs
- Includes ***any other pollutant*** "subject to regulation" under the CAA

Major Source Threshold

- For example, fossil fuels contain small percentages of sulfur, which is the source of SO_2
- Fossil fuels are mostly carbon-based organics
- PTE may be adjusted for practicably enforceable control requirements
- CO_2 controls?

Significance Threshold

40 C.F.R. Section 51.166(b)(23)

- Specifies significance thresholds for 11 pollutants
- For a "regulated NSR pollutant" not specified on the list of 11, "any emission rate" is significant

What Do We Do?

- Try a PAL
- Actual-to-future actual emission increase calculation
- 10-year look back
- Demand growth exclusion
- Be conservative about "routine"
- Follow legislative and rulemaking developments

Robert E. Holden

(504) 556-4130 Direct

REHolden@Liskow.com

LISKOW&LEWIS

A Professional Law Corporation

New Orleans | Lafayette | Houston

94b Clean Water Act Section 404: Jurisdictional Determinations and Nationwide Permits

Karen Janise Blakemore, Litigation Section, Environmental Group, Phelps Dunbar, LLP, 445 North Boulevard, Suite 701, Baton Rouge, LA 70821

Clean Water Act Section 404: Jurisdictional Determinations and Nationwide Permits

Summary of Presentation

The presentation will begin with a brief introduction of Section 404 of the Clean Water Act (CWA) then move into a general overview of the controversy related to jurisdictional determinations made by the United States Army Corps of Engineers (Corps). A discussion of the Rapanos decision will be included, as well as the jurisprudential attempts to clarify the Rapanos decision and the current status of the Corps' guidance.

Additionally, a general overview of Nationwide Permits (NWPs) will be provided which will consist of a discussion of modifications recently made to existing NWPs and General Conditions, as well as a summary of new NWPs and General Conditions.

Issues Addressed

Section 404 of the Clean Water Act (CWA) regulates the discharge of dredged or fill material into navigable waters. The United States Army Corps of Engineers (Corps) has the responsibility of administering Section 404. The Corps' authority includes determining whether the proposed activity will have potentially significant impacts to navigable waters. One facet of this evaluation is the jurisdictional determination of whether the proposed activity will impact navigable waters. Another facet is determining whether the activity itself will have significant adverse effects.

Jurisdictional Determinations

Although the term "navigable waters" was defined within the statute as "waters of the United States, including the territorial seas," this definition has not proved to be of much help in determining what Congress meant to provide the Corps in the way of jurisdiction.

Defining "waters of the United States" was a task left to the Corps, to be carried out through notice and comment rulemaking. When the Corps published its definition of "waters of the United States," it did so with the intent of following what it believed to be a Congressional mandate to give the term the broadest constitutional interpretation.

Current regulations define "waters of the United States" to include interstate wetlands, tributaries to wetlands, and wetlands that are adjacent to otherwise included waters and tributaries. The Corps further defined the term "adjacent wetlands" to be wetlands separated from other waters of the United States by man-made dikes or barriers, natural river berms, beach dunes and the like. This extension of Corps jurisdiction was the source of much angst for property owners and consequently, much litigation.

In 2006, the United States Supreme Court heard oral arguments in the consolidated cases of John A. Rapanos v. United States and June Carabell v. United States Army Corps of Engineer (herein referred to as Rapanos). Rapanos provided the Supreme Court with an opportunity to settle the disputes involving the Corps' jurisdiction over "adjacent wetlands" as defined in the regulations.

Much to the disappointment of those watching and waiting for a resolution to the issue of the Corps' jurisdiction, no such resolution was forthcoming. The decision split was 4-1-4, leaving the question of whether to follow the standard set forth by the plurality or the concurrence.

The standard presented by the plurality was that only those wetlands with a continuous surface connection to "waters of the United States" in their own right, are adjacent to such waters and covered by the CWA. The decision states that "waters of the United States" include only those relatively permanent, standing or continuously flowing bodies of water and that there should be no clear demarcation between such waters and those wetlands.

Justice Kennedy concurred in the judgment to remand, but rejected the reasoning set forth by the plurality. The standard presented in the concurrence requires the establishment of a "significant nexus" on a case-by-case basis when the Corps seeks to regulate wetlands based on adjacency to nonnavigable tributaries.

Simply put, the Court failed to resolve the issue of Congressional intent when it granted the Corps jurisdiction of the Section 404 permitting program through the CWA.

The fallout of the Rapanos decision is unsettled as Circuit Courts attempt to apply the elusive Rapanos rule and the Corps attempts to provide guidance for implementing the Supreme Court's decision. The Ninth Circuit, for example, applied the narrowest common ground of reasoning in Kennedy's concurrence. The First Circuit, on the other hand, was not able to identify any common ground and held that the Corps should define its jurisdiction by applying the standard set out by the plurality or the concurrence.

To ensure that Section 404 jurisdictional determinations were consistent with the Rapanos decision, the U.S. Environmental Protection Agency (EPA) and the Corps issued guidance to the Corps District Offices implementing Section 404 permitting program.

Determination of Adverse Impacts

In addition to making a jurisdictional determination of whether the proposed activity will impact navigable waters, the Corps must also evaluate the potential effects of the proposed activity. If the proposed activity is determined to have potentially significant impacts on "waters of the United States," the applicant must obtain an individual permit.

If, however, the Corps determines that the proposed activity will result in only minimal adverse effects, then the applicant may comply with Section 404 by obtaining a general permit. General permits are issued on a nationwide basis and authorize activities that have minimal individual and cumulative adverse effects on navigable waters.

The Corps issues nationwide permits (NWPs) to simplify the review process and allow applicants to proceed with fewer delays. In addition to meeting the threshold requirements of a particular NWP, the activity must also comply with NWP General Conditions common to all activities as well as any applicable Regional Conditions designated by each Corps District Office.

In March of 2007, the Corps published a reissuance of NWPs. All existing NWPs and General Conditions remain in effect, though a number were issued with modifications. Additionally, there were six new NWPs and two new General Conditions.

94c Assessment of Vapor Intrusion into Structures on Property Involved in Real Estate Transactions: The New Astm Standard Practice

Erika Erikson, Guida, Slavich, & Flores, P.C., 750 N. St. Paul, Suite 200, Dallas, TX 75201

ASTM recently introduced a new Standard Practice for conducting vapor intrusion assessment at properties involved in real estate transactions. The tiered process is meant to supplement the existing Phase I environmental site assessment standard practice (ASTM - E1527), which does not explicitly include an assessment for vapor intrusion potential. The first two tiers are screening tiers designed to assess the potential for a a vapor intrusion condition to exist. if it is reasonably probable that a vapor intrusion condition may exist and it is desired to confirm the situation, the third tier in the standard presents a "toolbox" of actions that can accomplish this. The fourth tier addresses mitigation alternatives. It is not necessary to progress sequentially through each tier in the process. For example, from Tier 1 one can move directly to any of the other tiers or even directly to mitigation in Tier 4.

94d Katrina: Regulatory Lessons Learned

Mary Ellen Ternes, McAfee & Taft, 10th Floor, Two Leadership Square, Oklahoma City, OK 73102-7103

In the aftermath of Hurricane Katrina, while emergency responders and administrative agencies were overwhelmed with initial response efforts, law makers were also busy debating legislative proposals to address perceived gaps in existing federal law. In response to these proposals, the American Bar Association Section of Environment, Energy and Resources (ABA SEER) reviewed the emergency response issues identified by the CDC and the EPA and the reactionary legislative proposals arising from the events, from broad exemptions to expedited cleanups, all in the context of the pre-existing regulatory and statutory environmental exemptions. ABA SEER produced a White Paper summarizing the existing exemptions and regulatory flexibility and analyzing possible risks from the proposed statutory fixes, concluding that such legislative efforts were both unnecessary and would have created greater risks in an already chaotic time.

This presentation will summarize the events, the political response, ABA SEER's legal analysis including the more than 50 existing exemptions federal law currently provides to environmental requirements, and their conclusions for future use by regulators and regulated entities.

AUTHOR INDEX